Step by Step

The Story of a Tulip

It Starts with a Bulb

Lisa Owens

Lerner Publications ◆ Minneapolis

Lerner Publications Company
An imprint of Lerner Publishing Group, Inc.
241 First Avenue North
Minneapolis, MN 55401 USA

For reading levels and more information, look up this title at www.lernerbooks.com.

Image credits: Olya Humeniuk/Shutterstock.com, p. 3; nacroba/Shutterstock.com, pp. 5, 23; gorillaimages/Shutterstock.com, pp. 6–7, 23; Kanjanee Chaisin/Shutterstock.com, p. 9; jessicahyde/Shutterstock.com, pp. 10-11; Abigail's photos/Shutterstock.com, pp. 13, 23; Olexandr Panchenko/Shutterstock.com, pp. 14–15; VladyslaV_Portfolio/Shutterstock.com, pp. 16–17, 23; Andrew Fletcher/Shutterstock.com, p. 19; Sakdaporn Chinjaroen/Shutterstock.com, pp. 20–21; Viktoriia Hnatiuk/Shutterstock.com, p. 22; Cover images: Elena Zajchikova/Shutterstock.com; New Africa/Shutterstock.com.

Main body text set in Mikado a Medium.
Typeface provided by HVD Fonts.

Library of Congress Cataloging-in-Publication Data

Names: Owens, L. L., author.
Title: The story of a tulip : it starts with a bulb / Lisa Owens.
Description: Minneapolis : Lerner Publications, [2022] | Series: Step by step | Includes bibliographical references and index. | Audience: Ages 4–8 | Audience: Grades K–1 | Summary: "From a tiny seed to a sure sign of spring, there are many steps in the life cycle of a tulip. Close-up photos and straightforward text help readers discover the process"— Provided by publisher.
Identifiers: LCCN 2021000219 (print) | LCCN 2021000220 (ebook) | ISBN 9781728428246 (library binding) | ISBN 9781728431659 (paperback) | ISBN 9781728430881 (ebook)
Subjects: LCSH: Tulips—Life cycles—Juvenile literature.
Classification: LCC QK495.L72 O94 2022 (print) | LCC QK495.L72 (ebook) | DDC 584/.69—dc23

LC record available at https://lccn.loc.gov/2021000219
LC ebook record available at https://lccn.loc.gov/2021000220

Manufactured in the United States of America
1-49361-49465-4/2/2021

Tulips are beautiful.
How do they grow?

First, gardeners buy tulip bulbs.

They plant the bulbs in the fall.

Next, they water the bulbs.

Then gardeners wait for spring.

In spring, the tulips sprout.

Then the
tulips grow.

Finally, the tulips bloom!

After blooming, the tulips store energy.

The next year,
they bloom again.

Tulips can bloom for years!

Picture Glossary

bloom

bulb

gardener

sprout

Learn More

Gates, Margo. *In the Garden.* Minneapolis: Lerner Publications, 2020.

Koestler-Grack, Rachel A. *Bulb to Tulip.* Minneapolis: Bellwether, 2020.

Nelson, Robin. *The Story of Honey: It Starts with a Flower.* Minneapolis: Lerner Publications, 2021.

Index